TALKING PLANTS

Me!

What's
Up?

SUSAN SCHADT PRESS

susanschadtpress.com

by Joyce Markovics

Published in the United States of America by Susan Schadt Press
New Orleans, Louisiana

www.susanschadtpress.com

Content Adviser: Dr. Kathryn Morris, Professor of Biology and Director of Undergraduate Research, Xavier University, Cincinnati, Ohio

Book Designer: Ed Morgan
Book Developer: Bowerbird Books

Photo Credits: cover, title page, table of contents, 4–5, 8–10, 19 top, 22, 24–29, 33, and 35, freepik.com; © Jonathan Bloom, 7; © Feri Fitrian Krismoko/Shutterstock, 11 top; Acroterion, Wikimedia Commons, 11 right; © Bildagentur Zoonar GmbH/Shutterstock, 12; © Pek Kool/Shutterstock, 13; © 5 second Studio/Shutterstock, 14; Public Domain/Wikimedia Commons, 15; Wallpaper/Adobe Stock, 16–17; © Thijs de Graaf/Shutterstock, 18; Alvesgaspar, Wikimedia Commons, 19 bottom; © ifong/Shutterstock, 20; © Tin-Kong/Shutterstock, 21; © Neelakandi/Shutterstock, 23.

Library of Congress Control Number: 2025900071

Printed in China

You're looking mighty sharp today, amigo!

CONTENTS

UMM . . . WHAT IS THAT?

What *is* that sound? It was a mic-drop moment for scientists in Israel. At first, they couldn't believe their ears. But after recording a group of plants, the researchers made a surprising discovery. They learned that wine grapes whine. Tobacco plants moan. And tomato plants complain. Plants, it seems, can "talk"! However, they don't chat like people. Plants communicate using clicking sounds that sound like popcorn popping or bursting bubble wrap.

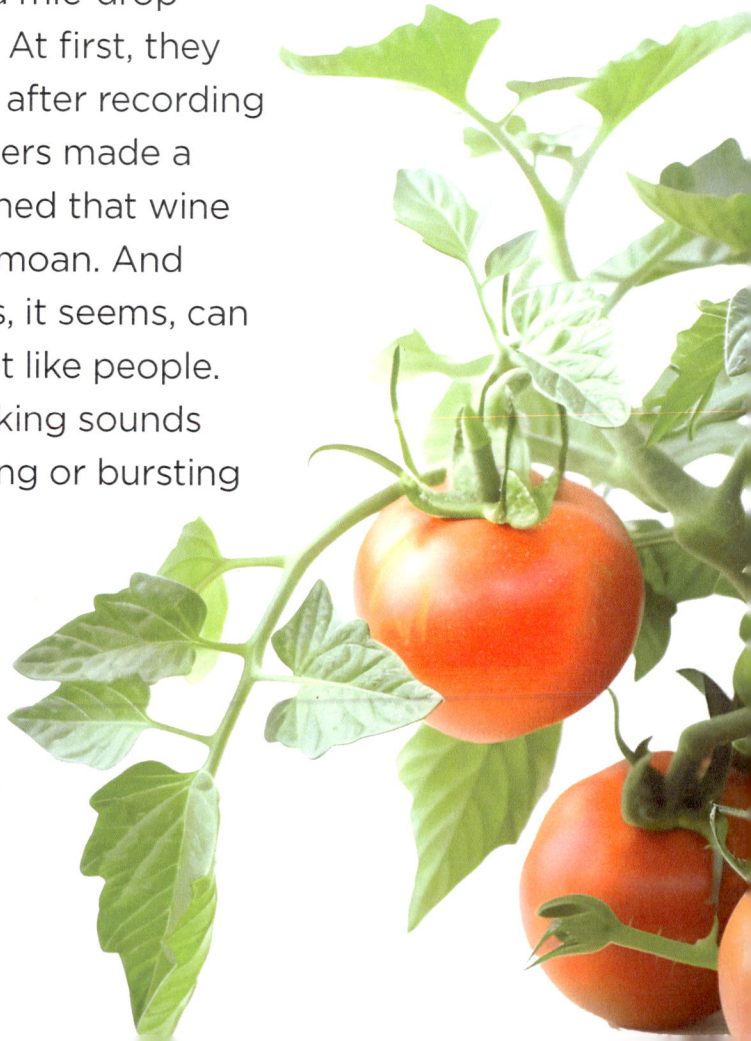

I love you from my head TO-MA-TOES!

Patterns soon appeared. The scientists learned that some plants make certain sounds for specific reasons. "When a tomato plant is feeling well, it **emits** very few sounds," said lead scientist Lilach Hadany. But that changes when it's stressed. For example, a plant clicks when its stem is cut or it doesn't get enough water.

A tomato plant's popping sounds can be heard up to 16 feet away!

However, the plants' clicking sounds were too high-pitched for human ears. So, Lilach and her team used special microphones to hear them. The team also wanted to be sure the plants were making the sounds. So, they recorded them in soundproof boxes.

Over time, the researchers realized that thirsty tomato plants make lots of noise. Stressed tobacco, wheat, grape, corn, and cacti are big talkers too. Incredibly, the scientists learned to tell the type of plant just from its clicks. They also used sound to identify the kind of stress the plant was under. In other words, "I'm thirsty" clicks are distinct from "I'm under attack" clicks. But, other than scientists, *who's* listening to these plant sounds?

PSST!

Researchers are still figuring out how the plants make the clicking sounds. One possibility is that bursting air bubbles inside the plants' stems cause the sounds. This is called cavitation (kav-ih-TEY-shuhn).

"Even in a quiet field, there are actually sounds we don't hear," says Lilach, pictured above. "And those sounds carry information."

WHO'S LISTENING?

"There are animals that can hear these sounds," Lilach said. Moths and mice, for instance. "Plants interact with insects and other animals all the time." Experts know that most animals communicate using sound. For that reason, it makes sense for plants to use sound too. Experts are also considering if plants can respond to the sound of other plants in distress.

In the future, farmers could benefit from talking plants. Picture a field of corn. Imagine if the stalks could tell a farmer that they need more water—or less. What about if they could signal they're under attack by pesky insects? There's still a great deal to learn. But one thing is for sure—plants are silent no more.

The most widely grown crop in the United States is corn. It's mainly used to feed livestock. Here a farmer examines corn kernels.

PSST!

Plants are key to life on Earth. They provide food for humans and other animals. Plants are also used for building materials, making cloth, medicine, and countless other things.

9

Duckweed is the world's smallest flowering plant. It has tiny green leaves and roots.

PHENOMENAL PLANTS

The clicking plant **experiment** raises some puzzling questions. For one, what do we really know about plants? Humans understand some basics. That is, most plants need air, water, nutrients, and sunlight—like humans do. We know that green plants, in particular, can make their own food. This process is called **photosynthesis**.

HANDS OFF!

Many other plants also seem to know when they're being touched. Researchers have shown that plants that are touched a lot don't grow as tall and produce thicker stems. Think twice, tree huggers.

The sensitive plant is also known as *Mimosa pudica*.
This means "shy" in Latin.

PSST!

How do some touch-sensitive plants move their leaves? When they're triggered, there's a decrease in water in their cells. This causes the leaves to fold up.

THINK GREEN!

Okay, so plants might not be able to talk like us. But they can communicate in ways we're just beginning to understand. From what we know, they're sensitive living things—and utterly unbe-LEAF-able! Just think, the next time you cut a flower, it might be screaming for help. Or consider what a **pungent** herb might be saying to a hungry animal.

Are plants intelligent? This question is still up for debate. Yet most experts agree that just because plants don't have brains doesn't mean they don't have their own form of intelligence. If your houseplant could talk, it might say, "Expand *your* mind, buddy!"

When grass is cut, it releases a chemical alarm signal. So much for enjoying the scent of a freshly mowed lawn!

36"

Eww! One of the largest—and smelliest—flowers is the stinking corpse lily.

Also, plants can't exactly move on their own from place to place. There's no such thing as a pole-vaulting pole bean! Even though they're rooted in one spot, plants have amazing abilities. They can still make or attract food, defend themselves, attack enemies, and **reproduce**. Scientists are learning more about plants every day—and having their minds blown.

PSST!

Hundreds of thousands of different plant species live on Earth. Experts say many more have yet to be discovered.

The tallest known tree is the coast redwood.

"I WILL SURVIVE."

It's clear that plants "know" some stuff. Or else they wouldn't have survived for 500 million years! They are great **adapters**. Plants can respond to ever-changing **environments**. Just think of what they have to deal with: Armies of pests. Droughts, floods, and fires. Competition from other plants. And the list goes on.

I'm so FROND of you!

Ferns first appeared on Earth 300 million years ago! They're one of the oldest plant groups. This tree fern forest is in New Zealand.

SPORES

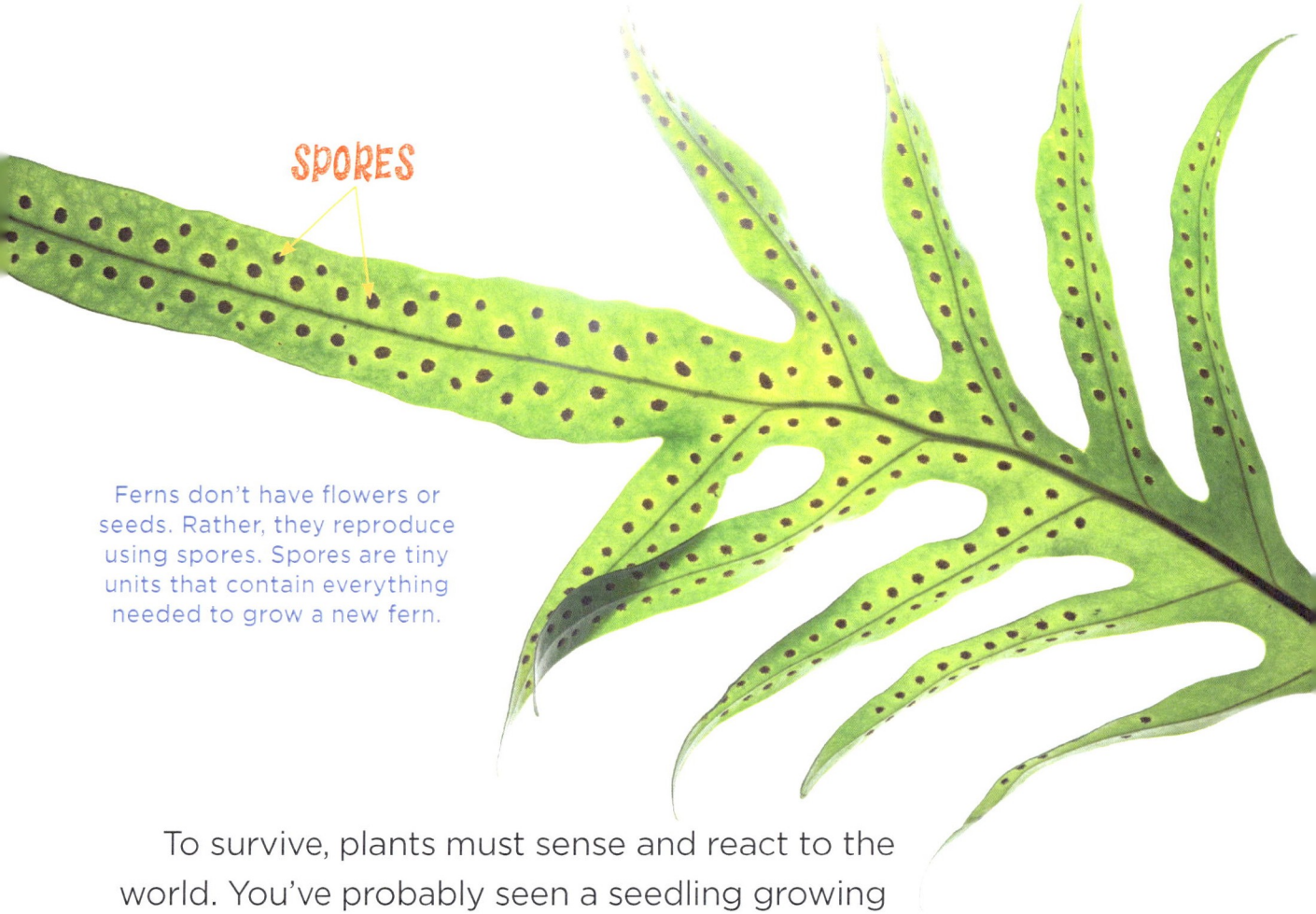

Ferns don't have flowers or seeds. Rather, they reproduce using spores. Spores are tiny units that contain everything needed to grow a new fern.

To survive, plants must sense and react to the world. You've probably seen a seedling growing toward the sun. Or picture a desert tree that has long roots to access water deep underground. These are a few basic ways plants stay alive—and thrive.

PSST!

Chomp, chomp! Did you know plants can lose up to 90 percent of their stems and leaves and still live? Picture yourself with only a head!

To flourish, plants have **evolved** superb senses. Humans have 5 senses. Plants have at least 15—and likely many more such as the ability to detect gravity. There are some interesting connections. But plant senses are very different from those of people.

My favorite rock band is the RED HOT CHILI PEPPERS.

14

BRAINY BEHAVIOR

Another difference is plants don't have brains like humans. English scientist Charles Darwin thought roots functioned as a plant's brain. He was wrong. Even so, plants can still sense and interact with their environments. But *how* exactly?

Charles Darwin was among the first scientists to believe that plants can communicate. Here he is as a young boy.

PSST! A 1973 book claimed that plants like rock and roll. It also said that plants can read our minds! These and other silly ideas were disproven. Still, the book was a bestseller.

15

SUPER SENSES

Plants don't have ears (except corn—*hee, hee*). But there's recent evidence that proves plants can "hear." Rather, they can pick up **vibrations**. In one study, scientists played a recording of munching caterpillars to a group of plants. A second plant group was played wind or other nature sounds. Or they were kept in silence.

Most caterpillars feed on leaves. Some plants make bad-tasting chemicals to keep caterpillars from eating them.

When real caterpillars attacked the first plant group, the plants made lots of chemicals to defend themselves. The chemicals came in "faster and often in greater amounts," said Heidi Appel, one of the study's lead scientists. The second group made less chemicals. She also **observed** that the plants could distinguish between recorded and real caterpillar, wind, and other sounds.

The plants knew when to defend themselves against actual threats. "That resulted in big grins," Heidi said. "Sound absolutely matters to plants."

PSST!

Have you heard of killer tomatoes? When attacked, some tomato plants release a chemical. It causes caterpillars to stop eating the plants. Then they start devouring each other!

LISTEN UP!

Can plants hear other things too? Another experiment, led by Lilach, showed that evening primrose plants are also good listeners. When her team played buzzing bee sounds to the flowers, they sweetened their **nectar**! Sweeter nectar attracts more bees—and perhaps leads to more **pollination**. That's one brilliant bloom!

Who says you need ears to hear? A bee feeds on an evening primrose flower.

These findings wowed Lilach. She also noticed the evening primrose has a specially shaped flower. It looks like a small satellite dish. This bowl shape is best for picking up bee sounds. "We found a **potential** hearing **organ**!" said Lilach.

PSST!

When researchers plucked some of a primrose's petals, it didn't pick up the bee sounds as well.

UNDERGROUND SOUNDS

Scientist Monica Gagliano wondered about what plants can "hear" underground. So, she planned an experiment with pea plants, soil, and recordings of water sounds. We "put the plant into a container which had two tubes at the base, giving it a choice of two directions for the growth of its roots," said Monica. Could the plant's roots pick up the recorded sound of running water in the soil?

I'm ROOTING for you!